BOG TREASURE

Eileen Carey

Eileen Casey and Jeanne Cannizzo

Bog Treasure

Bog Treasure

is published in 2021 by
ARLEN HOUSE
42 Grange Abbey Road
Baldoyle
Dublin 13
Ireland
arlenhouse@gmail.com
www.arlenhouse.ie

978–1–85132–266–4, *paperback*

Distributed internationally by
SYRACUSE UNIVERSITY PRESS
621 Skytop Road, Suite 110
Syracuse, NY 13244–5290
Phone: 315–443–5534
supress@syr.edu
syracuseuniversitypress.syr.edu

Typesetting by Arlen House

cover image:
'Frozen Sphagnum Boghole'
by Tina Claffey
is reproduced courtesy of the artist

Contents

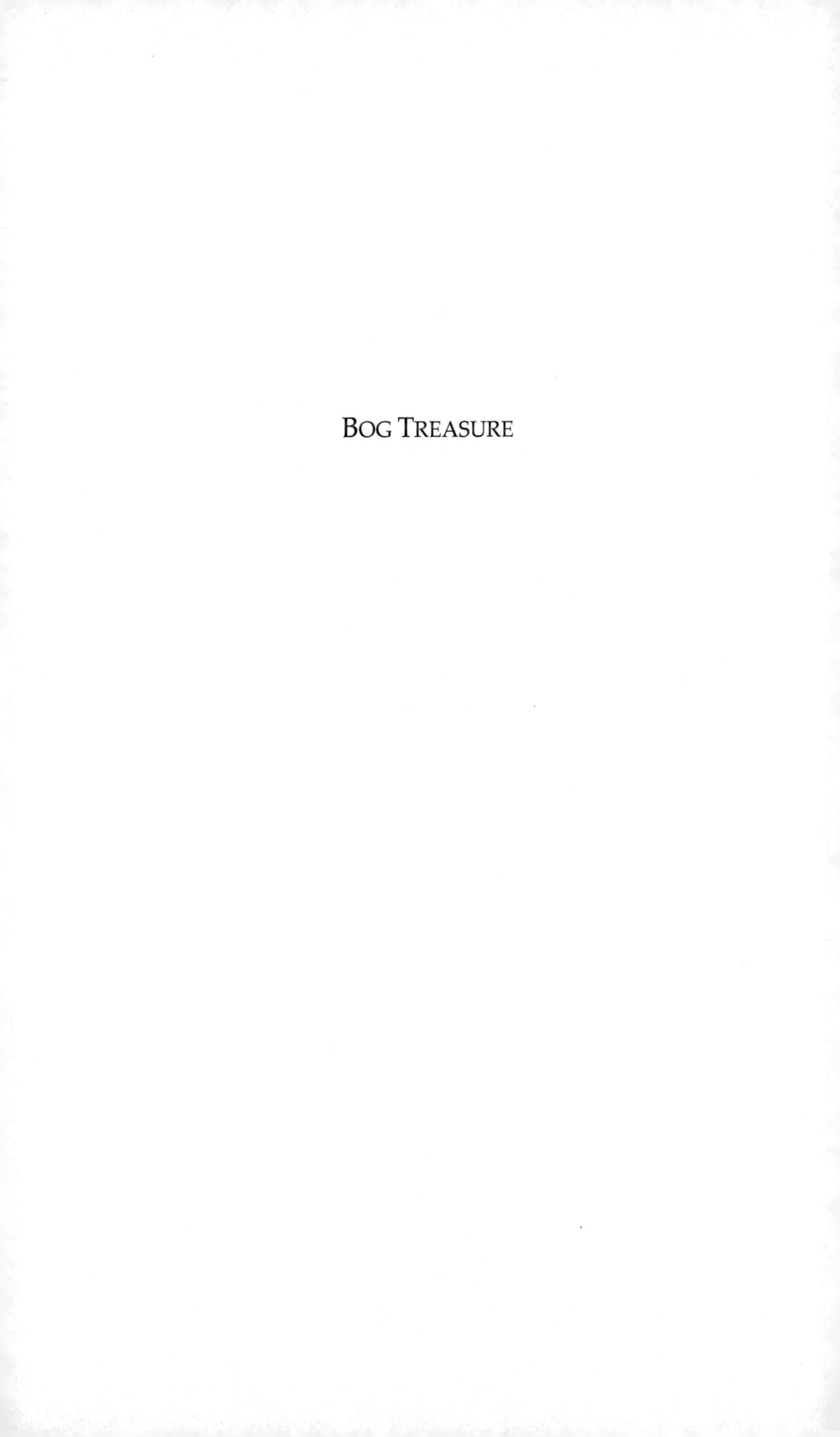

Bog Treasure

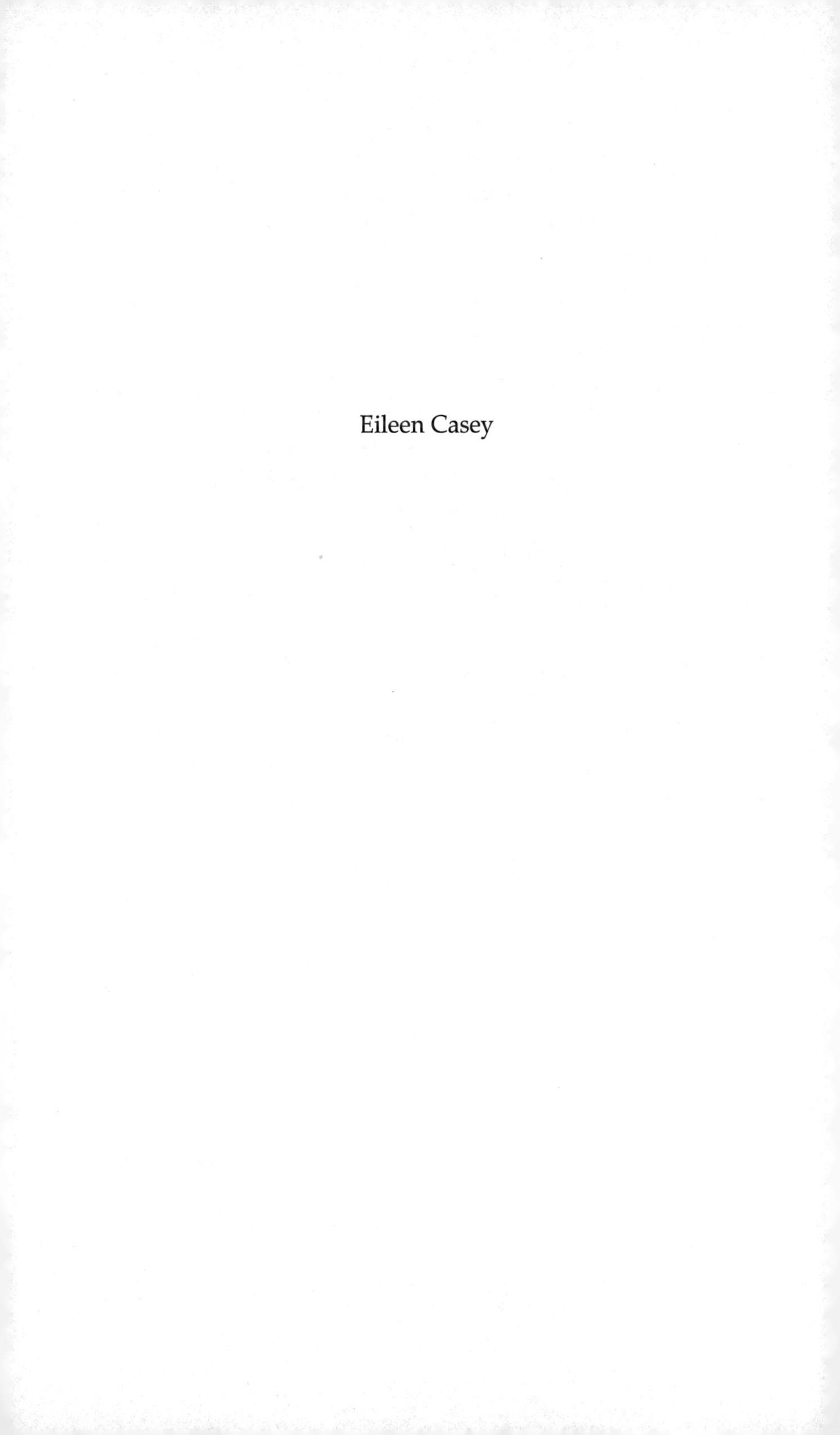

Eileen Casey

Eileen Casey, poet, short fiction writer and journalist, is originally from the midlands. Her work is widely published, in anthologies by Arlen House, Faber and Faber, Dedalus, New Island, *The Nordic Irish Studies Journal, Orbis, Abridged,* among others.

Berries for Singing Birds (Arlen House), her fifth poetry collection, was published in 2019.

A Hennessy/*Sunday Tribune* award winner for short fiction, she's also a Patrick and Katherine Kavanagh fellowship recipient. Recognition for her poetry includes among others the Oliver Goldsmith International Prize and a Hennessy/*Sunday Tribune* shortlisting.

As a mature student she received a BA (Humanities) from DCU (2009) and an M.Phil (Creative Writing) from Trinity College, Dublin (2011).

Poetry exhibitions include *Poetry on the Wall* (South Dublin Libraries, 2000), *Seagulls* (solo exhibition awarded by Tallaght Arts Centre, 2006), *Reading Fire, Writing Flame* (solo exhibition awarded by Offaly Arts, 2007) and *The Jane Austen Sewing Kit* (Birr Theatre & Arts Festival, 2009, Clare County Library, Ennis, 2010).

Previous collaborations include poetry/pen and ink drawings, poetry/shoe images, with County Offaly visual artist Emma Barone.

Artistic Statement

Bog and all things peaty, entered our midlands home via baskets of brown sods. Rectangular shaped, threadbare ends suggested an unravelling of sorts. Like an old but familiar rug coming undone. Still clinging to bog layers. And not without struggle. This latter proved in itself a recurring motif with regards to family emigration. Generation after generation. Leave takings but few homecomings. My brother Noel lives in South Carolina and often speaks of the 'Pocosins' (evergreen shrub bogs on the coastal plains of the south eastern United States, from Virginia to Florida; most common in North Carolina). 'Pocasin' is an Algonquin (native American) word and means 'swamp on a hill'.

Peat sods, on the Irish bogs of my childhood, were stacked in pyramids, for drying, or in a linear kebab shape, one sod on top of another and held in place by steel construction, again like a skewered kebab. Because bogs are swampy, it took a while for the sods to dry out enough to bring home, usually on a cart pulled by a donkey. Those sods were doused in the scent of bog rosemary and asphodel and steeped in birdsong; the skylark, golden plover and hen harrier. Mixed with logs or briquettes, they smouldered for hours, drifting various bog odours around the kitchen. Down the length of my street also. Those same wisps of bog scented smoke meant that neighbours were either cooking or boiling water for washing.

When I grew older, I'd pass by the bogs on bus rides to larger towns and counties. It was a bit like driving through the Nevada desert, towards Vegas glitter. On a smaller scale of course and without the scorching heat. More in terms of expectations; looking forward to bright light excitements. Sometimes bog cotton caught my eye. Snowy lanterns that waved soft as the bus roared by. In the city, because of my Midlands background, I was sometimes referred to as a 'bogger', an affectionate term I'm now, as I was then, proud of. I later discovered that what seems like a vast stretch of flat, uninteresting landscape, is a fertile breeding ground for thriving flora and fauna. Often surprising, like the delicate bell shaped bog pimpernel, soft sedges and heathers.

When Jeanne Cannizzo and I decided to collaborate on 'The Strange Case of the Irish Elk,' the stage was set to explore some of the many facets of bog lore, not least being the fact that discoveries of great elk bones have been found, mostly in Irish bogs. Deer or elk? Seamus Heaney wrote a series of bog poems, alert to its romantic (prairie like) possibilities as well as its ability to preserve for posterity. In 'Bogland', (*Door into the Dark*, Faber & Faber, 1969), Heaney lifts out a great elk from the bog and crates him in air. He uses the word 'elk' and by surrounding him in empty air, I believe Heaney gives the elk space to roam in new, unfettered territories. Our imaginations.

What draws me as a writer to the bog is the surface and depth elements, that sense of outward calm while underneath is written layer upon layer of narrative. Heaney described these layers as being like the pages of history. Collaborating with Jeanne Cannizzo is a delight. I find her work inspiring, her process nurturing. She is a true original with tremendous creative vision.

Dust

Sirocco winds sieve fine Sahara drifts onto
windscreens. Nature's exotic gift. Blood
thicker than water. Bog dust thicker than sand
shed from peaty pyramids. Summer after summer,
under fingernails, lodged into crevices. Back-
breaking graft. Cold tea slaked thirst. Butter
melted on bread. Lapwing or plover overhead.

Bog dust smells of wet winters. Even in heat.
Fathers, lone cowboys with donkeys. Head
Honchos. Their tower blocks rising sod by sod;
nearest thing to skyscrapers on bog prairie.

Mothers complained. Bog dust hidden in newly-
washed linen. Powdery clots crushed into seams,
rolled up sleeves. Shaken out on washboard ribs.

Flat bog prairie sometimes yielded secret bounty.
Discarded metal, old buttons. Bodies. Bog spirits
invisible, freed through trepanned skulls. A Spitfire
sunk in a Donegal bog, lifted out, all guns blazing.

If we'd only known how deep elk sleeps, deaf
to our vibrations. Thorny, like holly, his antlers
prized. Sold at auction to the highest bidder.
What rituals we'd perform to raise this Lazarus!
Dervish dances. Passion stirred in storms of dust.

Treasure

Dappled light pleats lilac shadings.
Blue meshes with pink; bog weathered
morning enters its stride. Colour
sharpens as light deepens. Spider webs
drape lacy antimacassars across purple
heathers, yellow flowered asphodel.
Early frost begins to thaw, burgeons
sphagnum's already swollen hoard.
Dew glistens pearly frogspawn,
dragonflies hover close by. Skylarks
rise with meadow pipits and willow
warblers or stall over a bog-bean pool.

Human and beast leave traces in their wake;
a thumbprint traced in buried bog butter,
a psalter creased by righteous devotion,
elk bone fragments. Bodies. Stabs of
bog shadow struggle with bog memory;
sacrificial wounds. We glimpse survival
in russet-edged leaves, mauve bruises
ruffled onto moss.

Bog is like a treasure filled galleon.
Centuries deep. Imperial measure in peat.
We lose sight how even inconsequentials
become more than their sum of parts.
Faithful to its seasons, bog keeps track.

CLONYCAVAN MAN

Ballivor, County Meath, 2003

Upper torso dredged to surface by harvester.
Treacly-skinned chest and arms, face flattened
into bog like a discarded leather satchel, forgotten
until turned by 21st century man. Squash-nosed.
Disembowelled. Axe felled. A life unfolds.

Death dated 392BC–201BC. Iron Age, last of
three epochs. Killing a king arrived in threes also.
Drowning, strangulation, bludgeoning. Celebrating
three goddess forms. South European pine resin oil

inched Clonycavan man's stature higher. Imported
hair gel spiked vanity. Reflected in a mirror,
portal to an Iron Age underworld. In this looking
glass he saw the last of himself. Forehead shaved,
hair piled high, glistened like fresh blood.
He could be a man from anywhere or everywhere,
off to meet his lover, grab a coffee or a beer.

Appearance is everything. Even then. Before ritual
blows crushed bone, split his skull in two.

Walckenaeira Alticeps Loves Sphagnum

Might be graffiti written by a love-crazed, moss-
glutted *Walckenaeira Alticeps*, though too
unwieldy for public wall or bus shelter.
On the eye, jelly soft, beetle body hard as shell.
Beady legs look peas in a pod translucent,
slime textured. Fat. Gorged on pinks, burnished
coppers. Drenched in bright green bog pools,
crawls over loose springy moss. Slippery
snorkel stems moist enough to satiate.

Cloncreen Wind Farm (2030)

Flocks of snowy gulls, wind-spun miracles;
far from seashore on Cloncreen Bog.
South East of Clonbullogue, Rhode,
towering turbines. Bog returns to green.
Blade rotations no longer puzzle warblers
or skylarks flying over wind propelled
energy. The same glorious gusts balloon
shirts on clotheslines like carnival swing
boats. Waltz dresses, skirts. Or polkas.

Bog comes into its own, what Aristotle
decreed for even the tiniest acorn. Bog destiny,
a carbon tank. No more peat cubes stacked
for winter burnings. Bog restored, gifts
healing, the yaw of just transition.

Naoscach

Hot, humid landscapes, Sierra Leone,
Senegal. Exchanged for Boora. Shy
observer, hidden in damp reeds. Dark
brown, black striped plumage bestows
camouflage. Long bill flute music heard
on a bog summer's evening. Fuelled
by drumming bleats, zig-zag flights
vibrate wing-spanned feathers. Naoscach.

Irish for snipe, scientific name *Gallinago,*
Gallinago. Echoes Astronomer
Galileo, Galileo.

You also belong to the night. Outlined
stark against skies scattered with stars;
each a mirror of bright bird eye.

HUNT THE SNIPE

An animal concocted from fancy
pranks us towards fool's errand
in forest or fen, mountain top
or furry glade. We search out
fantastical beasts; ear of ram,
neck of swan. Newt eye.

We plunder imagination's map,
know full well you are named for a bird.

The last of our truly magnificent perished.
Is this what sends us in search of you?

PEAT

Water soldier, bulrush, gypsy wort,
birch. Turf-brown shades,
fire smoke, fire smell.

Star moss, feather moss,
cowberry, crowberry,
bog blood.
Constant as rain pulses
over Pollagh, over Clara,
where rannock rushes grow.
Across Boora near Sliabh Bloom.

Yellow flower bog-bursts
flecks flat landscapes
of bog people.

Preservation in textured layers.
Pine, bog-rosemary, burgundy
red bog moss, water horsetail.

I am of the bog people,
fire smoke, fire smell.

Cranberry Bog

Pequot tribe, Cape Cod, name them 'ibimi';
bitter berries glutted by bears. Wiry stems
prove no match for dagger claws, grizzled
jaws. Related to blueberries, bilberries,

'Mad Sweeney', an Irish king, adored them.
Cast out and cursed by St Ronan, his travels
in the wilderness sweetened by fruit vines.

Runners snake along acidic bog surface,
ruby beacons. Mónóg,* natural dye, poultice
for fever. Blossom and stem chameleon,
resembling a crane's head. Below juice-
soaked surface, a bloody underworld
cloaked by crimson. Wounded landscape
battle scarred. Ancient warriors wild with red.

* *Irish for Cranberry*

Elk Disappearing Act

I imagine elk as flesh and blood man,
emerald clothed. Once verdant, his life
itinerant before climate vagaries bound
him in hunger's chains. Padlocked –
Houdini fashion – his future lay in a giant
acidic tank. Cuckoos almost extinct, hide
in counterfeit nests. Willow warblers cling
to reeds. Mindful of yet another pandemic
shift in a world suspended in time, I ponder
his vanishing; imagine him trussed in sedge.
Sunlight reaches his pinhole eye. *Camera
Obscura* shows him partial eclipses or
blind reality, guarded by lazy damselflies.

ELK SKY

Diana granted Orion place among celestial
plains. Elk also adorns the heavens. Horned
moons navigate night acres. Burnished
as the Bear's sword and belt. Slipped
behind rooftops, a hunter's moon.

Howling winds muffled by washing line
flaps. The night is alive with elk.
Shaped in ghostly outline curved
above my window, carbon copy traced
in antlered stars. Such skies warm winter
skin, guide me to dreaming realms.

Prometheus Reborn

Maddened by flame gifted an ungrateful world,
Zeus chained Prometheus to a cliff edge.
An eagle tears flesh from bone. Prometheus
re-births. Cursed to relive raw agonies.

Chained to our storyteller's rock, extinction
casts its narrative net. We haul up the pieces,
drowned fragments from lakes or bones buried
in bogs. We formulate new plotlines, new
possibilities. Strip it all back into 'what ifs?'

The Mesolithic Loop

Lough Boora Parklands

A tarred pathway in Boora, opposite
the bird-hide. Beyond the partridge farm,
coniferous forest. Remnants of a storm
shoreline exist, when River Shannon
and its lakes – Lough Ree and Lough Derg
– covered much of the midlands. Proof of
colonies, earlier lives marked in charcoal,
V-shaped axes. Made from hard stone.

Age to age, Mesolithic hunters transitioned,
survived by foraging. Sat by roasting fires
outside long timber houses. Eating fresh kill.
A Mesolithic barbeque of sorts. Or chewing
gum, made from birch bark. Hazel nuts
found in hazel woods. Children suckled,
then slept, sung to rest by Mesolithic
lullabies. Flame flickered across rosy skin,
star lanterns studded dark Mesolithic skies.

Whatever time or place, we leave traces.
Folds, creases, pinch-points. Our comings
and goings mapped out like scent drifted
from history's open windows.

The Mesolithic hazel trees, so like those
circled around the legend's sacred pool;
salmon feeding on leaves. Hazel wisdom
gifted to Fionn Mac Cumhaill. Heartache

also navigates its own trail, generation upon
generation. Desecrated forests. Ravaged
landscape. Birdsong silenced. There's joy
in preservation. Our future in safekeeping.

Bog Haiku

Star galaxies stir
egg-shaped feast of wonderment
moon spoons into bog

Acidic surface
brushes a hare's bristled skin,
paradoxical

Over mounds of peat
the whooper swans arc in flight
winter wetlands' birds

Cotton lights fleck bog
far as roving eye can see
summer white lanterns

Great Elk mysteries
birth ancient mythologies
noble beast fallen

Dripped into silence
once upon a wetland time
an ancient world stilled

Wide wooden pathway
village to village foot-bridged
roads diverged in bog

Winter brings herdsmen
wind shakes snow avalanches
undergrowth revealed

The Conversion of St Hubert

November fens wake to Hunter's Horn.
Breezes sulk on alder branches, a hunter's
moon slips its amber caul. Bog myrtle
drifts incense in scented ritual.

Old bones stir to life. Festered wounds bleed,
pierced by spears of senseless kills. Converted
on Saint Hubert's Day, hunter transfixed by
shine of silver crucifix hung between giant
stag antlers. In a casket of peat, reliquaries.

Birds of prey, elk spirits. Softened by nature's
sanctity. Bog-moss. Bog-flower.

The Strange Case of the Irish Elk

Elk. The word kicks at our throats
like fine whiskey. Irish. An Fia Mór.*
Centuries interred in bog, brought
to surface from peaty layers. Crated
in air by Heaney; freedom forges
new territories. If desecration brings
just reprisal, it's to see ebonised
magnificence picked clean to bone.
Worshipped by tribesmen, elk felt
Siberian winds, knew African horizons,
crossed Eurasia. Stranded in his frozen
world, felled by hunger's icy daggers;
why wouldn't we claim him as our own?

* *Irish for Great Elk*

Presence

Each time I pass a bog,
I feel him bloom.

He's in there somewhere,
cossetted centuries deep.
Layered through serrated gloom
perched on a ledge of rime.

Crowned with thorns
his universe not yet born;

he sank to his knees,
the mighty fallen.
Mired in his mythology.

Each time I pass a bog
his giant elk skeleton
gleams white as bog
cotton. 'Will-o-the-wisp'
lanterns flicker moonlit
spools. Like lotus flowers
rooted in murky pools.

Each time I pass a bog,
I hear such loud bugling calls
un-frozen from a dead mouth.

Marl

Mother came from Galway, sea blue in her eyes.
She settled in the midlands, a perfect ocean mirror.
Calm surface gave way to frolicking sea horses,
if angered or when she sang. Music in her blood.
Father's silence created its own sound. Rarely,
we knew the workings of his mind. Always cold,
he'd sit by the stove. Turf smell overpowered
wood used to stretch out its life expectancy.
A triangular turf dome, stored in the garden
stole light from an already dark interior.
Silent and dour to her gaiety, peaty brown
eyes stared into the distance, away from us,
that small kitchen, our two up, two down.
Layer upon layer of years, they stayed
together. Two different people. Sea and bog.
Yet, some bogs, near the surface, contain a shell
like substance, brittle yet enduring, reminder
how clay and water produce a living bog.

Fisher of Elk

Sieved from Lough Neagh, elk head, antlers.
Hauled on board by a fisherman; a prized,
unsalted catch. Strange to see elk netted
from lake, porthole eyes empty, algae slimed.
Not in Heaney's *crate of air** but drenched
by liquid pearls dripped onto boat floor.

* *'Bogland' by Seamus Heaney,* Door into the Dark *(Faber, 1969).*

Antler Lichen

Found in bogs. Wet northern forests.
Black spruce, white cedar, pine bark.
Courage totem for elk spirits, each
unhurried, one step at a time. Light
starved, raggedy-edged.
Packing for mummified bodies
cushioned into threadbare afterlives.
*The Ebers Papyrus** claims antlers,
applied externally, cures headache.
Osseous remains near Qantir,
seat of Ramesses, 19th dynasty,
date to Holocene Period.
Whether indigenous or imported?
Positives or negatives? Blurred
as an out of focus photograph.

* *Egyptian medical journal dating to 1550*

Mud Turf

Layer after layer cut into. Balayage
squares thrown to catcher on a lower
bank, passed to barrowman below.
Poetry in motion.

Deep shades. Deeper yet. Mud bedrock
black as Hades, turf gold. Burns fierce
as coal. Trousers rolled to knee, toes
splayed, bare feet squelch to harden.
Skin-dyed bog. A wild dance as when
threshers stamp juicy vine fruits
into luscious wine. Wayward flamencos.

Coming home, skies darken. Sphagnum
flickers prickle greenish hues. Bat silhouettes
stark against a wafer moon. Sacred notes
heard in even the tiniest tipples of night pools.
Bog trout in brackish waters swimming.

Black Forest, Boora Bog

Named for a forest in Bavaria, trees rooted
down sepulchral centuries. Green shoots
tannin dyed; moulded by darkness.
No nesting birds, no nestling moons
curved in crevices between bark and branch.

Stag God *Cernunnos* dipped his midnight
quill in raven's ink. Spiky as a porcupine,
silhouettes outlined. He created ferocious
dragons, fiery breaths frozen in charcoal.
Slithery serpents resilient in oak. Picasso
style. Raised from ghostly realms by human
hands, glorious as stalactite emperors.
Rain blesses elk antlers, feathered swans.
Whatever shape Eye of the Beholder sees,
sunshine gilds. Daylight rinses clean.

*Keep a green tree in your heart and perhaps
a singing bird will come.**

** Chinese Proverb*

Unlikely Adam

after seeing a tree sculpture of an elk by Andrew Carragher

You took branch ribs, twisted them muscular,
created a great elk. Rack of antlers delicate
by contrast, kindness bestowed perhaps
for such burden. Moulded from ash.
Light coloured, straight grained. Durable.
Smooth to touch. Alder, tree of Bran,
God of prophesy. Shields forged courage,
seagull sanctuary. Hawk. Raven. Outer tones
conceal darker clotted orange sap, as if blood
mingled. Sycamore's inter-locked grain,
memory of leaves, five-pointed stars.
Imported from as far away as Asia,
North America, places great elk knew.
Coming to our shores across ice-melted
land bridges. And so it was you breathed
life into branch bone. Adam without Eve.

Ready to bolt, stance rigid with tension. Head
flung back, startled by crackle of twig. Danger
spoors scenting the air. Like Lot's wife fleeing
Sodom and Gomorrah, unable to resist an urge
to look back, taking inventory of a world grown
smaller, possessed of fewer possibilities.

Siberian Ice Maiden

Made from larch, plentiful in Siberia,
your chamber door opened, fast-tracked
a Bronze Age world into a modern summer.
Resurrected old stories, already transformed
by myth. Reality scripts other realms.

Stories outlive their time. Subterranean
artefacts tell their tale. Two small tables,
horsemeat. A wooden vessel with carved
handle. Beverage in a horned cup.

High Priestess, you might have stepped onto
pages of *Vogue*. Dressed in a yellow silk
tussah blouse, crimson and white striped wool
skirt, thigh high white felt leggings. Cause of
death, also modern. Cannabis to quell pain.
In life. And death. On the boundary of steppe
and mountain, on either foot of the Urals.

Mummified, your body belongs to another
epoch. Elaborate Sintashta burials. Horses,
chariots, deer. Riding together into afterlife,
'The Pastures of Heaven'. Chieftains buried
with sacrificial animals, as you were. Horses
masked and head-dressed to look like deer,
antlers fashioned from wood or cloth.
Markings tattooed on your skin, likeness
on polished mirrors, tell giant deer were gods
worshipped in Ukok.* Herdsmen bring sheep
and horses to this plateau during winter's want.
Wind blows so fierce, snow falls off grazing grass
tiny avalanches revealing lush green undergrowth.

* *Ukok is now The Altai*

Moss Miracle

Wet-slimed, like seaweed gathered on seashores,
sphagnum's antiseptic bandage staunched wounds,
helped soldiers ruptured to bone survive
mud filled, rat infested trenches.

Sphagnum's memory bank records changes
stored in branch and stem, leaf and bark.
Swallow clusters flown over lakes thick
with frogspawn. Wetlands pierced by curlew's
plaintive cries. Autumn sunshine blazing on
heather. The hen harrier in motion. Orion's
Belt. Taurus. The Plough. Aching clarity
in skies untroubled by fluorescent mortar explosions.

Sphagnum absorbs up to twenty times its weight,
water or blood. As warriors from myth knew well.
Battle scars cool. No cure for deeper wounds.

Bog Train at Boora

We need more night for the sky,
More blue for the daylight
– John Ashbery

Iron ribbed vehicle of light pitched against the sky,
space for all weathers. Sullen winter at ease
with swallow flits; motionless air wing sliced.

Once upon a Boora time, bog train rode bog plains,
peat cargos packed in metal pockets, held in place
by steel rivets, heavy with purpose. Spring pokes
a tease of wild bog rosemary through slats
in metal floors, like ivy climbing through deserted
windows in a house alive once with human voices.

Six carriages yoked to engine, caboose for tea,
sandwiches and craic. Men on a break, playing
cards, *Snap* or *Twenty-five*. Reading the paper or,
with a sidelong eye, each other's faces. Far
more reliable than ink. Spectral ghosts now.
Fireflies sparked in a maddening dance,
out of steam. Perhaps bog train transforms
to charabanc, time rusting back to bare frame.

Bog train. Receptacle for delicate cobwebs,
skeletal across this landscape. Or a bronco
bucked out of place. Back broken.
You also broke the back of countless labour.
Ferried the peat that lit the fires that stoked
engines that fed industries that brought work
which supported families in houses Jack built.

In dreams, a man rides his shunting journeys
on a bog train sheened by sunshine. Before
everything derails he leans out into July heat,

waves at a grey partridge or a scuttling hare.
Or at nothing or no one in particular. Leaving
a trial of thick, dense smoke signals in his wake.

The Jesus Christ Spider

Raft Spider's story is more biblical than myth.
The only one of its species doesn't spin webs,
complex as deceit. No delicate loops or arcs,
or narratives spun from delicate threads,
artistry woven by Arachne or Athena.
Raft spider is a hunter. His pincers, venom
filled, aquatic. He runs on water in quick
pursuit of prey. Four pairs of legs, swift
as an Olympian. Dark brown, striped white
either side, silver under water. Can eat
small frogs, sticklebacks, tadpoles.
Attracted by the female's long slender legs,
all four pairs, he performs a bobbling dance.
He has no wings, no antennae, nothing
but his wits. Mating over, he disappears.

Woman Wearing her Home Around her Shoulders

Where you live means 'end of the world'.
A mammoth calf, thousands of years old,
found on this peninsula was one winter young
at time of death – the same age as your baby son.

No tree-lines shield. Tundra winds unpick
your way of life, loosen stitching on your Yamel home –
a tent made from dried reindeer.
Your husband drank its still-warm blood,

coming to you all those months ago,
raw flesh in his mouth, smells of slaughter
sown into each crease and crevice of his skin.
Below the mark where steel struck bone,

together at night you lie. Steel slit the reindeer's throat
wide as the opening where your head pokes out.
You gaze upon these rolling miles and it seems
as if your home is wrapped around your shoulders.

You gauge the time to rise like birds, make your way
from north to summer pastures in the south;
waiting for the river Ob to freeze (already late)
while all along Siberia's northwest coast
thousands of barrels are emptying out your world.

WITNESS
after a charcoal sketch by Jeanne Cannizzo

Spectral Modigliani limbs, murky
past elongated to present tense.
Lean flesh masks depths hidden
in hollowed eyes. Mouth a
charcoal twitch, a bristle broom.
Sweeping away layers of toxic
grime. So we can see stars.
Portals straddle out of proportion
spaces. There's a road map here,
destiny's twists and turns.

Paved promises, nowhere
pathways. Distances floundered
as we ride the horns of antlered
dilemmas. Naked ambition perks
us up. Despite famines of soul.
Disease ridden universes
preserved in rain-sodden
trenches. Yet and yet again,
caught in gleams of moon
a different light blinks back.
Fraught with possibility, beyond
flesh and blood. Metaphysical
transport in a brown cube of peat.
I press down hard upon the *sléan*.*

* *bog shovel*

LOT 441 (FOUND POEM)

A Pair of Irish Giant Deer or 'Irish Elk' Antlers.
Circa 10,500–8,000BC.
House Sale, 24–25 September 1984 (*Christie's catalogue*)

Lot Essay

Most likely, Mervyn Wingfield (d. 1904),
7th Viscount Powerscourt acquired these giant deer
or 'Irish elk' antlers.
Christopher Hussey, in his 1946 article described
the 7th Viscount's collection scarcely matched.
Of ancient and notable stags' horns ...
and, of course, an Irish elk head,
one of the finest specimens of this extinct giant.

Created by Richard Wingfield (d. 1751).
1st Viscount Powerscourt, from 1728–43,
to designs by Richard Castle. From 1842,
its formal gardens laid out by Daniel Robertson,
in the grand Baroque manner
of princely gardens in Austria.

Price realised, £52,875.
For a specimen with fifteen points, skull restorations.
Five points with restored breaks, one antler with restored
break near the skull. Excavated from a bog near Limerick.

It's a long way from desolate bog
to Christie's bustling saleroom.
Sound-proofed slumber exchanged for hard cash.
Trophy for the wealthy. A solid investment.
It doesn't hurt to have Powerscourt provenance.
Titanic salvage for the curious.

Antlers at Hatfield Hall, Hertfordshire

Diggers leave mangled bone, cartilage, fibrous
tissue. Retrieved by hand-held implements,
un-damaged antlers turn up in sale rooms.
Photographed, a pretty girl's face often screens
elk skull. Giant antlers sprout a set of wings.
She's like a *Victoria Secrets* model.
Marketing. Sex sells. Even fully clothed.

Anne Boleyn visited Hatfield before Henry's
passion cooled; his roving eyes settling on Jane.
Sister to Thomas, a man Elizabeth I, at 15,
schemed to marry. Treason seasoned word,
gesture. Her nerve held, fiery as her hair.

Elizabeth's silk stockings and gloves, first
of their kind in England, display at Hatfield.
Doused in history, like antlers sourced
from a beast long extinct. Hung beside
sumptuous brocades, steeped in blood
and betrayal. Tudor legacies. Bog sacrifices.

Bog Wish

Bury me in sound-proofed soft sedge.
Let me lie protected by antlered wings.
No shroud. Instead, fold me in summer
wild bog flowers. Keep every orifice free
so bog juices fill up every pore, so
my story flows for centuries to come.

Plant no tree. Create no sculpture as mark.
Leave nothing between me and naked sky;
bog birds will know. I'll hear skylark's
song. Iridescent ibis will shimmer all
across this dark, dank landscape.

So I, an ordinary woman, contribute
layer by carbonised layer. Sealed in bog
rosemary, asphodel or heather. Blood
of my bog blood floating in myriad
coloured sphagnum beds.

MAY TIME BEE

I find you, hovering over dazzled light,
faithful to your choisya flower. Long
bee tongue, mandibles moving, you
gather sweet nectar. Black and orange
stripes blend white and lemon shades.
Drunk on vanilla scent, pollen pellets
swell on thread thin hind-leg. Wings
mirror translucent drips of slow time.

Come September, taken from warm apiary,
brought to chill bog air, you'll gorge on
ling heather. Some will perish. Others
carry on, busy among soft purples.

Returned to hive late autumn, bloated.
Port-wine-coloured honey in a jar. Row
after row, stagnant on supermarket shelf.

Mating Dance

Antlered branches brush against woodland's
low foliage. Earth scraped by hooves. A dying
man's surrender scrawled in clay. Watched
by gentle doe eyes, elk mating choreography
slow but showy. Molyneux believed

providence resurrects. Fossils coming back
into fashion like platform shoes or bell-bottoms;
bushy sideburns. Small revolutions confusing
elk with moose or reindeer. Curvier argues no.

Not so. Fossil remains remain fossils. Linked
to glacial un-peelings. Great Elk dance done.
Scattered to extinction by ego evolutions,
so Curvier says. Artful a tale as the style

worn by Suebi tribesmen.* Hair held back
in a thick knot. On battlefields, warriors
appeared taller, fiercer. It distinguished freemen
from slaves, separated the men from the boys.

** The Suebian knot found on 'Osterby Man' (70–220AD) at Osterby near Rendsburg-Eckenforde Schleswig-Holstein, Germany.*

Locking Horns

Two young bulls in a clearing,
battle ripe. Not for breeding
rights or territories they're goaded.
Threats, name calling, grudged
generation to generation. Hand
picked by tribal elders, knuckles
bare, blade sharp. Around the edges

small boys mime fathers, brothers.
Shoulders hunched, head to head,
legs sprawling back, footholds
grasped in dust. Short trousers
yet hard men in the making.
Antlers thicken on such drama.

The strongest bull lands the knock-
out blow. Stitches needed. Surgery.
Sometimes a hearse. Death better
prized than wounded family pride.
Phones flash. Chants and cheers.
A *YouTube* video soon goes viral.

A slow slip of time pools into pond. Becomes lake.
Nourished by hazel, alder, birch, rowan; layer upon
leaching layer transforms to bog. Peace drops low
over wetland reeds. Curlews write winter across
a landscape in mourning. Home for shape shifting

hare. Long-eared sentinel alive in legends. Spring
rain blesses rushes. Bog world strobes summer's
filmic gold, russet glints in bog rain mirrors.

Orange wings in Boora, near Sliabh Bloom
Mountains. Butterflies skating air. Hoof prints
traced in peat. Red deer stealth scarce visible.
Under moonlit plains; what ifs? For lost lovers.

November nights of the dead month, antlers
in wispy cloud shapes. Black eyes illuminate
bleak stirrings. Nocturnal skitters. Night lands
alive with bog myth. Orpheus searched dark

places for Eurydice. Thin spaces between light
and shade. Looking back, she was lost to him;
curious nature punished for love's cruel curiosity.

Spring Night

after Lia Bai (700–762)

From whose house are the flute notes floating?
Like autumn leaves in dry September's mulch.
They blend into the spring breeze all over Luoyang;
across this bog asleep in a cradle of stars
until dew-dank morning wakes music memory
played soft on membranes thin as a seasonal moment
merged into hours, days, weeks.

In the tune I heard expectant hands break willow twigs.
How a startled breath blows on bog perimeters,
gathers dusky dregs across spongy surfaces.
Scraped into the apron of cavernous lake,
knee-deep in lives no longer ebbed or flowed yet
felt in the hush of breaking light, settled darkness.

Who wouldn't think of home at such a moment?

LIBRETTO

Trapped on an island stage,
elk in a leading tenor role,
romantic as Rodolfo.
No baritone shortage,
frustrated mezzo-sopranos.
Already spring birthing
females bloat in chorus.

Soft pink-wombed baby,
membrane cushioned.
Glacial weather threatens
your debut. Soundtracked
by bellowing shrieks
across wasteland,
forest thinned.
No encore.

Two Birds, One Stone

A neighbour's dog is brought for burial. Old
sheets conceal his glistening coat. Its dark
shade names him a word odious now
as it surely was then. Innocents, we guessed

it somehow connected to colour.
Mrs H in her drapery shop described
gloves, shoes, handbags and coats
as such. No one batted an eye.

Newly dead, this Labrador dog put to bog
killed two birds with one stone.
He sank into dog heaven, his master loading
saved turf, winter's fuel stacked high.

Half a century later, I read a find in Lundby,
South Zealand. Elk remains wrapped in furs,
dating back 10,000 years. Thirteen sets
at least. Bones. Antlers put to rest with rite,
ritual. Ancients believing animals possess
spirits sacred bogs resurrect. Beloved.

Named in ignorance, poisonous as sundew.
Bog dissolves history, yet preserves its bones.

Midnight Melody

Bog breathing sphagnum, slow flickers
under a Rubens moon. Voluptuous frogs,
spawn-full, croak to each other across marshes.
Sacred music bless these midnight hours,

this luscious feast of darkness. Gorged
on by skitters of nocturnal creatures.
Insect ripples prickle surface and depth.
There is no sleep, only echoes.
Bat or bird. Wings open like flirtatious
fans. Swans conjoin in heart-shaped
silhouettes. Winter's harshest score
ices over; swells with prospects of thaw.

Lemanagh Bog Body

Between Ballycumber and Ferbane, surrounded
by River Brosna, Lemanaghan island bogs;
music composed for voices. Corhill, Tumbeagh.

Near St Manchan's Holy Well, Lemanaghan bog yields
a man's lower torso, mysteries in peat, unsolved over
two millennia. Flexed feet suggests force. Dragged
to burial after torture. Halved. Like a nation split in two.

His story could be today's headlines. Bound, gagged,
brought to an outhouse, hedgerow screened,
soundproofed. Clues to his whereabouts obscure as tyre
tracks on a dusty road. None but quarrelsome crows
see or hear cruelties perpetrated at his execution.

Blackthorn, stone axe heads; Corhill bog finds. Silver
coins hoarded at Curraghalassa, date back to Edward I.
Bog weathered, medieval shoes, shaped by warm flesh,
also Curraghalassa. Parish of Ferbane in the Barony of Birr.

Until torsos reunite, Lemanaghan bog body mystery
remains. Pilgrims visit St Manchan's site. Some believe
he restored life to his beloved cow.*
Stolen then slaughtered.
Then, as now, the milk of human kindness soured.

** St Manchan's cow was prolific in her production of milk. So much so that greedy individuals stole and slaughtered her. When the saint caught up with the thieves, the cow was already boiling in the pot. He put the pieces together and restored life to the dead animal.*

Time Capsule

Fossils watermark acidic depths. Layers
brought to surface. Rites, rituals, lure
storytellers, record broken lives.
Black gold spilt its hoard at Silkeborg.
Spongy, middle of nowhere desolation,
ochre grass sparse as matted hair.
Tollund Man. Tea-coloured skin.
Foetal. Roped neck. From bog tree
cut down or sacrificed where he lay?
Noose grown tight, breath taut,
ebbed to the quick like candle flame.

If he could plug into a cryogenic screen,
memory played out as in Potter's
Cold Lazarus. We'd see a man like any.
Father. Brother. Son. Lover. Mindful
of his season's end. Before a spade
scraped through to skull, daylight
trickling into his winter world.

Or Haraldskjaer Woman. From her bog
grave taken, shrouded in moss. Age, height
recorded. Unhusked millet, blackberries, last
meal. A neck groove, familiar appeasement.
Born out of bog god mysteries.

On show for all to see, in her princess glass
sarcophagus. Iron Age Snow White. No prince
has come to wake her, kiss her wizened lips.

JEANNE CANNIZZO

Jeanne Cannizzo is an anthropologist, artist, curator and sometimes a poet. A new poem features in the Spring 2021 issue of *J Journal,* from the John Jay College of Criminal Justice, New York. Her sculpture was shown in *Neoneanderthals,* for which she also wrote the catalogue and wall text, at the Royal Scottish Academy in Edinburgh, 2019. She lived in that city for several years while teaching at the University of Edinburgh and acting as an occasional guest curator for the National Portrait Gallery of Scotland. Now living in Canada, she has continued to exhibit her own work and act as curator for a number of exhibitions. In 2015 she put together *A Study in Contrasts* which featured works by Sybil Andrews and Gwenda Morgan, who both trained at the Grosvenor School of Modern Art in London, for the Art Gallery of Greater Victoria. For the Special Collections and Archives Library at the University of Victoria, she curated, in 2017, *From Dijit to Bridjit: A selection of letters from Sir William Orpen to Beatrice Elvery* (Lady Glenavy) and two years later, *Celebrity before Photography,* featuring 18th and 19th century engravings of theatrical portraits. Two of her collages with words were exhibited as part of the Arts Council of Victoria city-wide celebration of poetry, 2020. One of her paintings was the cover art for a special edition of *Adanna,* an American literary journal.

Artistic Statement

The first Giant Irish Elk I saw was in Edinburgh, at the Royal Museum of Scotland. It entered the collection in 1820; the most 'complete skeleton ever found'. Discovered in 1819 in a marl pit, on the Isle of Man, by a local blacksmith, it is sometimes called the Ballaugh Elk. Seeing that the specimen was missing its pelvic bones, Thomas Kewish is said to have fitted in those of a horse. Working with James Taubman, who was the tenant on the land where the find was made, they exhibited the skeleton for a fee.

Rumours of the find reached the Duke of Athol, the Queen's representative on the island. Through his mother he had manorial rights, and decided to assert them to gain possession of the splendid creature. Unwilling to surrender the skeleton, Kewish spirited it across the Irish Sea to England. However, he surrendered his prize when the duke won a lawsuit. Eventually it was presented by the new owner to what was then the collection of the University of Edinburgh.

I am not sure when I became aware of bogs. They are widespread in North America and I have been both attracted to them, and slightly repulsed, since I spied one in some wilderness – scraggy, black-watered, mossed deep. Some can actually support the weight of a moose on their spongy surfaces, allowing it to nibble at plants.

As an anthropologist, I am also intrigued by the idea that bogs are not quite land and not quite water either, but rather in some sort of liminal state – they exist betwixt and between them. This makes bogs ideal as ritual sites – there is a wonderful array of objects which have been thrown, lost, or deposited in them over many generations. There are also, of course, the famous bog bodies, about which I first read as an undergraduate in archaeologist Peter Glob's book on the mummies found in Danish peat lands. I have never forgotten the cover, featuring the haunting face of the Tollund Man, still wearing his sheepskin cap after being executed thousands of years earlier.

When Eileen and I were thinking of a project to work on together, the Giant Irish Elk, and those people and things found in bogs with it, seemed perfect. I am profoundly grateful to Eileen for giving me the courage to think I might write poetry. My husband, David Stafford, is always my reader of first resort and a steadfast supporter of all my new ventures – thank you.

The Art of Bog Swimming

A hat of felt floats on the bog's surface.
Has someone drowned? An accident?
Suicide, a nefarious murder even?
A ferret-faced head surfaces under the hat.
The bog swimmer is small-bodied,
slight but powerful. Back into still water,
to emerge shrouding himself with black mud.
Twisted, contorting ritual.

Two men watch. One begins to photograph
the twisted, contorting shaman at his ritual.
For this is Joseph Beuys, and this is art.
Performance art, needing documentation,
preservation for the unborn generations.
A commentary on the environment,
danced lament for disappearing peat lands.

Cave Lion Kills Giant Elk

That's the headline from *The Prehistoric Times.*
Not really news, I should think,
as far as predator and prey go.
But the detail is missing as to how she, the cave lion, did it.
How would I do it, if I were the killer?
The elk is gigantic, with huge antlers, so
I would have to sneak up on him. Or would it be better
to let him get a whiff of me, and run himself silly?
Then go for the throat, a frontal assault,
wrestle him down onto the ground to bleed out.
But all that blood smell, what else would come running
to get in my way or even try to turn me into prey?
It might be safer to come up behind him,
my soft pads not making a sound in the grassland.
And then with a single bound I launch myself
through the air onto his haunch, one side of the other.
Much better. Then I don't have to see those sorrowful eyes.
My cubs have to eat too, you know. End of story.

BOG BUTTER

'Butter to eat with their hog, was seven years buried in a bog.'
James Farewell, *The Irish Hudibras* (1689), English poet, lawyer

Cylinders of white – marled and mottled
by centuries of submersion in a bog.
Sculpted in dairy fat, packed and
pushed into a girdle of blackened wood,
wrapped in the bladder of a deer.

Bronze Age to Iron Age, unto the medieval
and even modern times, the Irish have buried
butter in their bogs. Weeks, months, years,
millennia in that frigid, acid water, no oxygen,
no bacterial growth, no mould, no rancidity.

Another Irish barbarism, a unique
backwardness, the English thought this
custom quaint. They knew nothing. An ancient
technique, many are the peoples who have
buried their food to store it without spoilage.

Buried, like golden hoards or a cache of weapons,
butter is a mysterious substance.
Born as liquid, transformed
by human hand in a process not fully understood,
it becomes solid,
liable to melt into something not quite either.
Like a bog, it is liminal in its nature.

So why not surround it with rituals and meaning
beyond nourishment? Make it a food reserved for the rich,
consumption governed by sumptuary laws,
pay taxes with it,
heal the sick, mark boundaries or make a votive offering.
Still edible after thousands of years,
surely bog butter is magical?

Born Still

I was never of their world.
The bog holds me; I am of this place.
Suspended in its colloidal grasp,
in the warm thickness of the dark,
I am comforted.

Dainty Toes

composed while observing the skeleton of a Giant Irish Elk

toes buttress the weight
keep the bones upright
when laden with flesh

Doe at Chauvet

The doe is alone, without her mate or herd.
Elongated neck, the tell-tale hump on the withers.
Legs almost too delicate to bear
the true weight of the enormous belly.
The womb seems already occupied
with the next generation.

Endling

The last of its kind,
unique among those dying animals.
The final one, a survivor, alone.
Waiting for its own death.
Herd animal with no herd.
Twin without its other half.
A remnant without issue.

And what if it becomes not a remnant but a revenant?
Cloned in an alien womb,
born into a new world, our world.
But still alone, without a herd?
Once the last is now the first.
De-extinction in an unfamiliar landscape,
strange predators, new diseases,
better, surely, as an endling.

INVENTORY – BOG 316

Reason for deposit

1 gold 9ct Claddagh ring, 17th century

Broken vows

1 platinum ring, small diamond, c.1935

Breach of contract

1,217 silver coins, Roman

Hidden from invaders

1 base medal coin, Euro 10ct

Accidental loss

1 silver sixpence

Discarded with pudding

1 stone axehead, prehistoric

Flaw in manufacture

1 bronze sword, 860BC

Ritual offering to supernatural

1 iron ring handled knife 3rd century BC

Discarded during fight

1 automatic pistol, C96 Mauser

Disposal during pursuit

1 blackthorn shillelagh, 18th century

Thrown at frightening shadow

1 fetus, male, 8 weeks approx

Spontaneous miscarriage

8 fetuses, 5 female, 3 male, variable

Abortion

1 newborn, male, 2 days old

Illegitimate birth

Make Mine an Irish Elk

Prayer at the tattoo parlour:
'Give me its animal power, its bravery.
Let me share its status, its magical beauty.'

Mostly on the bodies of men,
a corporeal identity
needled into the skin.

Often on the chest or forearm,
sites of masculine strength;
body culture in the marking.

The ink yields an embodied art.
Gender, age and ethnicity
expressed, but also formed by the act.

A visible boundary, a separation.
Some have the tattoo, and they are us;
an exclusionary ideology.

But the tattoo also signals a
community, one of shared values;
a place to be, a supportive herd.

Master Mind

Specialty Subject: The Bog Lemming

1. In what continent are bog lemmings primarily found?
2. Are the two lemming species identified as East/West or North/South?
3. How many teats does the Southern variety have?
4. How many teats does the Northern variety have?
5. Do bog lemmings hibernate?
6. How many teeth does the Synaptomys Borealis have?
7. How many toes does the Synaptomys Cooperi have on its hind feet?
8. What colour are the bog lemmings' droppings?
9. (Bonus) Name the Disney movie of 1958 in which a mass lemming suicide is staged.

MOLYNEUX

A kind face, learned, handsome
so Roubiliac's marble reveals.

A doctor with an army of appreciative patients
and a witches' cauldron of ideas.

Shakespeare's eye of newt to
Molyneux's swarming insects.

Toe of frog to Giant's Causeway.
Howlet's wing to Greek lyre.

Tooth of wolf to horns of giant elk.
Scale of dragon to Danish fort.

Salt-sea shark to elephant's jaw.
Baboon's blood to calculus.

Finger of birth-strangled babe to Irish taxes.
Root of night gathered hemlock to coughs and colds.

Out of the witches' cauldron come imprecations,
and foul visions.
But in Molyneux's simmer science,
healing and human curiosity.

NOMENCLATURE

I'm a deer. It's very clear.
Let me explain it all, right here.

'Irish Moose-deer' first said Molyneux of my bones.
Then Rutland too called his bull beast a 'moose-deer'.
And what of the 'elk' part?
Europeans describe as an elk (*Alces alces).*
What Americans call a 'moose' (*Alces americanus*).
A *wapiti* is the North American elk (*Cervus canadensis).*

Actually, we are all deer (Cervidae).
The elk, the moose and me.

PREDATORS

'The stomach rules the world. The great ones eat the less,
the less the lesser still.'
(William Buckland, 19th century geologist and cleric,
scientific observer of fossils in Kirkland Cave)

Think of Kirkland Cave, North Yorkshire, as a stomach.
What would a forensic analysis of its contents reveal?
A humble-jumble of splintered bones, broken skulls.
The last meal:
jawbone of giant deer;
forepaw of cave lion;
incisor of hippopotamus;
molar of rhinoceros.

Poor Buckland, wanting to maintain faith
in the Great Flood
but knowing that the fossils must be antediluvian;
prehistoric den for generations of cave hyenas,
floor littered with the bony remains of their kills.
For here they were the alpha predators.

Recipe – Haunch of Roast Elk Venison

First catch your elk.
You can:
dig a deep pit, cover it with grasses,
wait for the animal to fall into it. Or
take a sturdy net, stretch it between two trees
on opposite sides of a game trail, and drive the elk
into the net as it approaches. Or
spear the elk, making sure to sharpen
your spearheads in advance. Or
using as many hunters as you can gather,
release flights of arrows at the beast.
Failing all this, if you are near a cliff, take note
of the prevailing winds and hide yourself
at the edge of the forest. When a grazing buck or doe
appears, rise up shouting and waving your arms
to make yourself appear bigger,
and run it over the edge.

Now butcher the carcass.
Crack open the leg bones to reach the marrow; set aside.
Squash as many bog berries as you can carry
and grind them into a paste.
Rub this paste into the fleshy haunch.
Roast over a fire, using forked branches
of yet green trees to make a spit.
Baste with bog berry juices; do not overcook.
Serves an average cave full.

Household hints:
Scrape the hide as clean as possible, soak it in your urine
for 3 days to soften it and use as a cave mouth covering.
Surplus bones can be used for making awls.

Rituals

Someone is coming, the bog stops singing.
No slip-slapping of mud,
no gasping bubbles floating upwards,
no fetid gas hissing to escape.

The underlings gather to ascend,
to welcome the new one to the realm of peat.
But a noise without scale, too voluminous,
rents the body of the bog.

Wrong-doer or self-killed? Born still?
Captive or slave? Who is coming?
The corpse is hanging from
a monstrous cloth, suspended on the water.

Remembering the land of pain,
they do not wish to encounter the living.
With bladed iron they cut the cloth cords
and with gentle stealth they pull him down.

Boots, sinking past, are recognised.
Yes, a leather cap limned with wool,
chest protected by pelted armour.
A fighting man?

No judgments are rendered.
Every son, every daughter is honoured,
for those below were formerly above.
But still, is he a king's son? So many grave goods.

The underlings gather the un-nameable pieces of metal.
Bolts and rivets, the canopy, one propeller, a machine gun.
He must have been sacrificed,
for the people of the kingdom.
Above them all, the parachute darkens in the peat water.

Tattoos for an Irish Elk

My tattoos are my life, my autobiography.
Here a trefoil shamrock low on my narrow ankle,
in green, for Ireland, my motherland.
This broken arrow, red ink streaming on my withers
reminds me I survived the huntsman and his dogs.
On my haunch, a doe, a memorial to my favourite.
My chest carries a tattoo of the skull
and antlers of my kind,
trophies of rutting season victories in the past.
This is my self, inscribed on my body.

Buck at Lascaux

A prehistoric *Monarch of the Glen,*
such as Landseer never imagined.
On the walls of Lascaux – no need for canvas.
Red and yellow ochre, charcoal only.
Seeing the images, Picasso barely breathes.
'They invented everything.'

Clearly not a reindeer, who also
stalk the walls in herds, ever alert.
But a Giant Irish Elk, *Megaloceros Giganteus,*
with shoulder hump and ruffed neck,
palmate antlers so obvious.
Spectral bellows reverberating around the chamber.

THE DOE SPEAKS

Yeah, I know. It's the antlers.
But I would still like to point out:
I'm a giant too.
I'm just as Irish, or not, as he is.
And I'm an elk too, or not.
But I'm not in the museum hall of fame.
School kids don't 'ooh' and 'aah' while
gazing up at my skeleton.
Yeah, I know it's the horns.
Or maybe too few of us have been found.
Does, I mean.
Maybe that's it.
We're not so stupid as to fall into a bog.

The Kaiser's Antlers

Guess what the Kaiser got for a Christmas present in 1913?
Probably lots of stuff, but how about a pair of antlers?
Not any old antlers, but those of that noble beast,
the ancient Irish Elk.
Came from a bog near Dublin,
12 feet wide that particular rack.
Put them on the top of his Christmas tree,
the Kaiser did.
Not himself of course, but one of the servants.
Maybe those wide spreading antlers,
one of each side of the skull cap
a bit like a hand with the tines for fingers,
looked like the outspread wings of a Prussian angel
up there on the fir tree.

And guess who gave it to him? Deutsche Bank.
This is a long tradition. No, not the bank and the Kaiser,
although that probably is a relationship of some years.
The presentation of those giant antlers,
that's the tradition here, almost since the first pair,
monstrous and mysterious, were found.
From male to male, hunter to huntsman,
aristocrat to king, landowner to gentry,
and once in awhile, gentleman to museum.
Emblem of status, by birth or acquisition.
Attribute of power and virility, achieved or envisioned.
On display for all to acknowledge
the superiority of owner and elk alike.

The Pass of Plumes, 1599

Forward the forlorn hope.
Down the sunken road, entering the pass.
Five hundred yards, rebel trench at the end.
But they, the insurrectionists, are nowhere to be seen.

Forty men with guns, twenty short swords,
hardly more than knives.
Caliversmen to hold fire until their weapons
scorch the rebel chests.
Robert, Earl of Essex, Lord Lieutenant of Ireland,
is on the march.
Dublin to Munster will open his southern campaign.

Scrub woods on the embankments;
thickets hide the 'rogues'
and 'bare-legged beggars', as Essex calls them.
The road is rough, wet, no good for baggage,
a terror for horses.
Then the rear guard is caught
by Owny MacRory O'More and his allies.

Fighting pours out of the defile and onto the marshy land.
Great numbers, claim the Irish, fell
– 500 of the invaders lie dead.
So many, that the helmet plumes of Essex's dead knights
and the horses' bridle feathers cover the battlefield.

English sources make little mention of this first encounter.
Two or three officers killed, a few soldiers lost,
say the chronicles.
In the end, only the thirsty 'little bloody bog' survives,
triumphs.
Only it knows the truth of what happened
at the Pass of Plumes.

THE PEAT RETREATS

The peat retreats, the aerial heat increases,
the earth warms.
Each cycle unrelenting feeds upon,
fuels the other.
The earth warms, the aerial heat increases,
the peat retreats.
Drying, the bog surrenders
its cargo of dangerous gas.
The peat no longer takes that burden unto itself.

The disappeared return, yellowed skins now
exposed to the wind.
Achieving the bog's surface, the missing find
no familiar route home.
Those sacrificed escape their bondage,
some of the mysteries resolved.
But new unknowns emerge
as the dying of the peat,
and the destruction of the bog overtake us.

THE PRIVATE LIVES OF ESSEX AND ELIZABETH*

Quiet on the set.
Lights.
Camera.
Action.

ELIZABETH DAVIS *gazing out a Hampton Court window. Enter* ERROL ESSEX, *but the Queen doesn't turn towards him.*

ELIZABETH DAVIS (*angrily*): You have lost me Ireland, Essex.

ERROL ESSEX (*smirking*): Your Majesty, my Queen, my love, on the contrary, I have brought you Ireland.

ELIZABETH DAVIS (*turns and gasps as she sees Essex has attached a pair of giant Irish elk antlers to his head*): You fool, you stupid boy. I should never have let you take my army to that bog-riddled wasteland. How dare you present yourself to me with those ridiculous things on your head.

ERROL ESSEX (*taken aback that she doesn't see the joke but keeps trying*): Oh, come now, Bess. Look upon this gift I have brought you. 'Tis the rutting season, these huge horns are just for you, my sweet. (*Begins trying to do an Irish jig but antlers unbalance him and he topples over, then lying flat on his back, he grins up at the Queen*).

ELIZABETH DAVIS (*still not amused, puts one foot on his chest*): No more than a dumb animal, you are my Lord Essex.

ERROL ESSEX (*still playful, grabs her foot*): No, not so dumb, your Majesty. (*Gives out a massive bellow*).

Guard armed with pike and a lady-in-waiting rush into the room, bewildered. Essex has untied and removed the antlers from his head.

ELIZABETH DAVIS (*still angry but softening*): Take him to the Tower, before this jester, who loses Ireland and yet gifts it still, charms me again. (*Turns to the lady-in-waiting*).

Get someone to take these disgusting objects to the Horn Gallery, where I and all posterity can see Essex's fall from my grace and reflect upon his 'conquest' of Ireland.

* *With apologies to Bette Davis, Errol Flynn and Warner Brothers whose 1939 movie,* The Private Lives of Elizabeth and Essex, *won many awards.*

The Thrice Killed

Why have they done this?
Stabbed me, taken my head,
torn my body in two.

The axe in my chest brought death.
There was no need to kill me
again and then again.

Do I frighten them so much,
have I failed them so badly
they gather at the watery bog and perform
this hate-filled, bodily execration?

Do they feel liberated somehow?
Secure in the next harvest?
Safe now I am thrice killed
and will never return?

Underworld

The surface water of the bog is cool on the skin.
Sun sweetens the brown stains of the peat.
But that water is thick, not flowing,
not salty but not fresh either.
That is the nature of a bog – neither one nor the other.
A betwixt and between world – not land, not water.

Both and neither, a perfect site for rituals,
a theatre to stage the performance of the sacred,
enactment of death and for the door to the underworld.
The passage deepens, becomes sunless, colder.
Clot of moss, stick of laurel,
pitcher plant with captive spider
hang suspended in the glutinous vinegar soup.

At the bottom of the bog there is only stygian black.
If this be the entrance to the underworld,
so said the ancients, there will be greeters,
waiting for arrivals.
Grief, sadness, anxiety, panic, fear, forgetfulness,
pain, disease, old age, starvation, hunger, hate,
the sound of wailing.
How knew those ancients
of Ireland in its singularity and universality?

Zombie Elk

Bodies fight to the surface,
no longer bog-bound.

Viscera expand,
gas propelled upwards.

Exploding organs emerge.
Water graves give up the cadavers

spewing bacterial juices,
bloodied fur achieves the earth above.

Upon whom should the herd
take revenge for extinction?

ACKNOWLEDGEMENTS

The support of The Arts Council/An Chomhairle Ealaíon (Project Development Award) and Offaly Arts is greatly appreciated.

Sincere thanks is due to publisher Alan Hayes, Arlen House.

To errant artspace (Alston Street, Victoria, BC, Canada), for providing a venue for the work and for their willingness to embrace the collaboration between Cannizzo and Casey.

To Rita Ann Higgins, editor of *Out the Clara Road* (Offaly County Council, 1999) where 'Peat' first appeared and to AltEnTs (Rua Red Arts Centre, Tallaght, South Dublin) where 'Woman Wearing her Home Around her Shoulders' was first published in *From Bone to Blossom* (2008).

To photographer Tina Claffey for her cover image, 'Frozen Sphagnum Boghole'.